Brilliant

What the critics have said about *Brilliant!*

Theatrical magic. . . . This is a group to watch.

—*Globe and Mail* (Toronto)

Brilliant! is an amazing show.

—*Vancouver Sun*

For sheer intelligence and visual theatrical wizardry, it's hard to beat *Brilliant!*

—*Evening Standard* (London)

Brilliant! is.

—*Edmonton Journal*

One of this year's best shows. . . . The young artists of the Electric Company display a dizzying theatrical vocabulary.

—*Georgia Straight* (Vancouver)

With considerable charm and an abundance of invention, the Electric Company's perfectly counterpointed quartet of actors bounce ideas off each other in dizzyingly abstract fashion.

—*The Herald* (Edinburgh)

Richly imaginative.

—*Westender* (Vancouver)

Once upon a time, Hollywood used to give us popular movies about famous scientists and inventors. . . . The genre has long been out of fashion, but Vancouver's Electric Company revives it on stage, splendidly, with *Brilliant!*. . . Entertaining and enlightening.

—*Calgary Herald*

This is an all-out orgy of sight and sound.

—*Vancouver Courier*

In the hands of Vancouver's Electric Company . . . the inventor's tale becomes a Promethean fable about how visionary idealism is compromised by commerce. . . . As verbally assured as it is physically dextrous.

—*The List* (Edinburgh)

Theatre at its best seizes an audience and does not let go, not after they have left the theatre, not when they turn their own lights down. So it is with the richly innovative *Brilliant!*

—*The Daily News* (Kamloops)

Brilliant!

The Blinding
Enlightenment
of Nikola Tesla

Electric Company Theatre
Kim Collier
David Hudgins
Kevin Kerr
Jonathon Young

BRINDLE & GLASS

For more information on The Electric Company, or information on stage production rights visit www.electriccompanytheatre.com

Library and Archives Canada Cataloguing in Publication
Brilliant! : the blinding enlightenment of Nikola Tesla / Kim Collier ... [et al.].

A play.
ISBN 0-9732481-9-X

1. Tesla, Nikola, 1856-1943--Drama. I. Collier, Kim, 1965-

PS8600.B74 2004 C812'.6 C2004-906196-8

Cover and interior photos: Tim Matheson

Author's acknowledgements: Laurie Anderson, Mitch Anderson, Pat Armstrong, Susan Bartsch, Karen Berkhout, Brent Calkin, Linda Chinfen, Stuart Collier, Bill Costin, Courtney Dobbie, Wendy Gorling, John Guest, Rory Gylander, Christine Hackman, Jane Heyman, Jan Hodgson, Tim Howard, Carol Hudgins, Jamie King, Tim Licata, Duncan Low and the Vancouver East Cultural Centre, Kirsten McGhie, Una Memisevic, Paul Moniz de Sa, Adrian Muir, Playwrights Theatre Centre, Michelle Porter, Heather Redfern, Cindy Reid, Cindy Shaw, Kathryn Shaw, Margaret Tom-Wing, Samara Van Nostrand, Lew Wetherell, Kirsten Williamson, Azra Young, Canada Council for the Arts, British Columbia Arts Council, Department of Foreign Affairs, Province of British Columbia, City of Vancouver, Vancouver Foundation, Canadian High Commission, Leon and Thea Koerner Foundation, Hamber Foundation, Melusine Foundation, Individual Supporters

Brindle & Glass Publishing
www.brindleandglass.com

Brindle & Glass is committed to protecting the environment and to the responsible use of natural resources. This book is printed on 100% post-consumer recycled and ancient forest-friendly paper. For more information please visit www.oldgrowthfree.com.

1 2 3 4 5 07 06 05 04

PRINTED AND BOUND IN CANADA

For Pat Armstrong

FOREWORD

Brilliant! began its life in 1996 as an entry in the Vancouver Fringe Festival. The four of us first met at Studio 58, Vancouver's premiere acting training program. *Brilliant!* was the inaugural production of our newly founded Electric Company Theatre. The show became the foundation for a unique collaboration and creation style that would produce a series of scripts and productions in the coming years.

Between 1996 and 2003 we revisited and reinvented *Brilliant!* five times. In various incarnations it was a forty-five-minute Fringe spectacle, a two-hour-and-twenty-minute epic tragedy, and eventually, in this published version, a ninety-minute physically and visually driven tale of obsession, genius, and the fragility of an ego defined by the impulse to create.

Electric Company Theatre has always treated each element of production—staging, design, and performance—as a key component in the creation and telling of a story. Therefore, the script includes references to theatrical conventions, video projections, and non-verbal performance that are as vital to the narrative as any of the words spoken by the characters.

We encourage the reader to join the collaboration by reading between the lines and imagining the piece in performance.

Kim Collier, David Hudgins, Kevin Kerr, Jonathon Young
Vancouver, British Columbia, Canada
September 2004

PRODUCTION HISTORY

Brilliant! The Blinding Enlightenment of Nikola Tesla was first produced by Electric Company Theatre at the Vancouver Fringe Festival in September, 1996 at the Firehall Arts Centre with the following cast and designers:

Nikola Tesla	Andy Thompson
Katherine	Kim Collier
Robert	Kevin Kerr
Thomas Edison	David Hudgins
Phil	Anthony F. Ingram
Direction	Electric Company
Set, Sound & Costume Design	Electric Company
Video Design	David Epp
Lighting Design	Heidi Lingren
Stage Management	Nicole Vieira

A subsequent full-length production was produced by Electric Company Theatre at the Roundhouse Performance Centre in 1998, winning five Jessie Richardson Awards for Outstanding Original Play, Outstanding Production, Set Design, Lighting Design, and Sound Design. This production featured:

Nikola Tesla	Jonathon Young
Katherine	Kim Collier
Robert	Kevin Kerr
Thomas Edison	David Hudgins
Phil	Judi Closkey
Direction	Electric Company with Conrad Alexandrowicz
Set Design	Craig Hall
Sound Design	Attila Clemann
Costume Design	Mara Gottler
Video Design	Robert McDonagh
Lighting Design	Adrian Muir
Stage Management	Kelly O

The version herein was premiered by Electric Company Theatre in Edinburgh, Scotland in 2003 with:

Nikola Tesla	Jonathon Young
Katherine	Kim Collier
Robert	Kevin Kerr
Narrator / Edison	David Hudgins
Direction	Electric Company with Stephane Kirkland
Set Design	Andreas Kahre
Sound Design	Electric Company
Costume Design	Mara Gottler
Video Design	Amos Hertzman
Lighting Design	Adrian Muir
Mask Design	Melody Anderson
Stage Management	Jan Hodgson

PRODUCTION NOTES

The "sphere" referred to in the script embodies Tesla's singular vision. On stage it serves as a projection surface, physical apparatus and movable set piece.

In the stage directions, all titles in Roman upper case are projections. Scene titles, in boldface, are not projected; they are simply guideposts for the players and the readers.

The spinning newspaper headlines referred to on page 56 are a low-fi method of evoking the classic film effect. An oversized newspaper front page is mounted on a simple apparatus which the actor revolves as he rushes toward the audience.

Note on punctuation and overlapping dialogue (cf. Caryl Churchill):
A "/" marks the point in a character's line where the next character begins speaking.

For example:

Tesla: How many frequencies did he demonstrate?

Robert: I don't think / he even understands the concept.

Katherine: None.

Here Katherine jumps in with her answer to Tesla's question midway through Robert's.

If the line of dialogue with the "/" doesn't end with any punctuation mark it means that character keeps speaking on top of the other character's line through to their next line without break.

For example:

Katherine: All ours. Forget Anne Morgan—she's a tart. / Give us a

Robert: Tart. Good God.

Katherine: private—she is too! Give us a private show tonight!

Here Robert reacts to the word "tart" with line as Katherine carries on with her line without break.

CHARACTERS

Narrator: A shadowy figure obsessed with reassembling history. He is a ghost living in the future, haunting the past; a storyteller who believes that Tesla's legacy is the saving grace of humanity.

Nikola Tesla: A visionary Serbian inventor, whose pioneering work in electricity at the turn of the twentieth century ushered in the modern age.

Robert Underwood-Johnson: Editor-in-Chief of New York's popular *Century Magazine*. A poet and academic.

Katherine Underwood-Johnson: Robert's wife and New York socialite. One of the few people who ever became close to Tesla.

FBI Agents (2)
White Pigeon
Edison's Showmen (2)
Waiter
J. P. Morgan
Reporters (3)

New Yorkers
Pigeons
Lab Assistants

Preshow

As the audience enters, a human-sized canvas sphere stands illuminated onstage. Projected onto it, pigeons flutter and flock. Sounds of their cooing and flapping fill the space.

Prologue

Darkness. NIKOLA TESLA'S APARTMENT 1943. *Tesla is revealed: an old man carrying a bundle of letters. He struggles to open one but the stack falls to the floor, ruffling a nearby Pigeon who perches beside him. Tesla stoops and picks up the stack, this time getting one open. The Narrator appears.*

Narrator: He is a strange picture. Clinging to letters he has, until now, left unopened. That is how I see him at the end. Alone with his pigeons clutching a packet of dusty mementos. What does he fear? Perhaps he wonders that himself. He decides to read them—and that is the last time I can see him: as he is remembering her.

In 1943, Nikola Tesla died. The FBI confiscated three hundred thousand documents, most of them unpatented inventions and plans, packed into his tiny New York flat. Where the US government has these documents today is unknown. Last reported sighting: 1943.

Lightning. Thunder.

The Raid

In the semi-darkness, two FBI Agents break down the door of Tesla's apartment and rush in with their flashlights blazing. They search the room for Tesla's files, stacking boxes from his closet onto a dolly. Something stirs behind them and they spin to face it. The White Pigeon, frightened by the glare, flies away, revealing another box. The Agents uncover it. A moment of elation. They storm out with the confiscated files.

Intense music underscores this sequence with a series of images of Tesla's patents and inventions that roll like an opening credit sequence to the play:

POLYPHASE ALTERNATING CURRENT 1883
EARTHQUAKE MACHINE 1887
RADIO TECHNOLOGY 1891
ROBOTICS 1893
X-RAY TECHNOLOGY 1894
BROADCAST POWER 1900
ELECTROMAGNETIC PULSE 1901
RADAR TECHNOLOGY 1915
PARTICLE BEAM WEAPONRY 1917
DEATH RAY 1943

As the music plays out, the title appears: BRILLIANT!

Lightning Walk

A bolt of lightning. Rain. Night. BUDAPEST 1882. *Tesla walks alone in the hills near Budapest. We hear his voice, just discernable in the shadows.*

Tesla: Here, Faustus, is thy world—a world!
Still dost thou ask, why in thy breast
The sick heart flutters ill at rest?
Why a dull sense of suffering
Deadens life's current at the spring?

From the darkness, the Narrator appears:

Narrator: It's 1882, in Budapest. A young man is walking. His thoughts drift to a favorite passage from *Faust* by Goethe, and he speaks it aloud.

Tesla: It is, it is the planet hour
Of thy own being; light, and power,
And fervour to the soul are given,
As proudly, it ascends its heaven
To ponder here, o'er spells and signs,
Symbolic letters, circles, lines,
Then ye, whom I feel floating near me,
Spirits, answer, ye who hear me!

Lightning flash. Tesla experiences a vision. As he draws it on the ground, the diagram appears.

Narrator: He freezes as if he is being clutched . . . from the inside. Collapsing to his knees he grabs a stick and draws in the sand. The diagram. The answer. It is a rotating magnetic field. It is the heart of the Polyphase Alternating Current System. A system that will transform the field of electricity, transform all of industry, transform all of society, transform the entire world. And he knows it.

Lightning flashes. Blackout.

Rowing

Night on the Hudson River. Sound of ocean waves. NEW YORK 1884. *Robert and Katherine are in a rowboat together.*

Robert: The Brooklyn Bridge is fading in the fog. Maybe we shouldn't be so far out. Are we taking on water?

He bails.

Katherine: Starlight, starbright, first star I see tonight . . . was made by Thomas Edison. One, two, three, four, five, six, seven, eight, nine, lights all together there. It's an electric constellation. Twinkling on the shore. I wonder if there is anything we humans can't do?

Robert: Do you see the particularly bright set over there?

Katherine: Yes.

Robert: That's the office of J. P. Morgan. I think he believes that light is not meant to see by but to be seen by. Oh, I should commission an editorial cartoon about that.

Katherine: What an arrogant turd.

Robert: Why?

Katherine: Not you. Morgan. He's offensively ostentatious.

Robert: He's rich.

Katherine: He's noisome, noxious, nauseous.

Robert: Yes, and he's invited us for dinner.

Katherine: Seriously? When? Who else will be there? Oh I can't wait to see the inside of his house!

Robert: How quickly the tide has turned.

Katherine: No, no. What better revenge on one I can't stand, but to eat his food, drink his wine, and charm my way onto his permanent guest list.

Robert: You ruthless villain!

Katherine: Look at our life! Who are we?

Robert: I don't know, but apparently J. P. Morgan does.

Katherine: I guess when you run a popular magazine those that crave being seen want to be seen by you.

Robert: And to think, Mother said that journalism was no place for the respectable.

Katherine: Well it isn't—that's why it's so fun! Can life get any more exciting? What are you looking at?

The biggest, loudest, lowest foghorn belches its warning directly behind them. Katherine spins around, only to see a colossus looming over them. They both scream. Tesla stands above, holding the rails of the ship. Surveying the harbour of New York, Tesla catches the eyes of the occupants of the small boat. Turbulence and confusion as the giant vessel scrapes alongside them.

Robert: God almighty!

Katherine: Happy anniversary, darling!

They are swept offstage. Katherine is laughing.

America!

Tesla stands at the prow of his ship. Images of the Statue of Liberty being constructed pass before him.

Narrator: Nikola Tesla arrived in New York the same year as the Statue of Liberty. But this colossus was barely distinguishable from the other immigrants. Carrying only a few coins, several poems, and a packet of obscure calculations, his letter of introduction set him apart. A message from a leading scientist addressed to a Mister Edison: "Dear Sir: I have met only two great men, and you are one of them. The other is this young man."

Thomas Alva Edison was thought to be the world's greatest inventor. But when Tesla's ship arrived in the harbour, Edison was unaware that he, his phonograph, and his light bulb would soon be yesterday's news.

The crackling strains of an early phonograph recording of "God Bless America" fill the air. Tesla climbs down the ship's ladder and into a flurry of city life. Customs officers frisk him and push him through a turnstile. Hucksters bedazzle him, while pickpockets go to work. Snake oil salesmen hawk their wares, halting him with their perfumes, notions, and gizmos. An opportunistic photographer hands him an American flag and snaps his picture. Before he can recover from the camera flash, he is swept up into the chaos of a marching band parade.

At the end of it all, Tesla, in private, composes a letter.

Tesla: Dear Mother:

Another letter I dash off between fitful daydreams . . .

America is everything we've heard it to be.

The day I arrived I met the man they call the World's greatest inventor: Thomas Edison. Within the hour I was working for him; a half-hour more, already finding flaws.

Now he's offered me fifty thousand dollars to improve some of his patents.

All seems possible here.

I need only convince him of the merits of my new system. Ours is a glorious time.

Aside from these details, there is little to report. But I long for the day when I shall see you again.

Yours,
Ever loving,
Nikola

He exits under the pounding, grinding, cacophonous sounds of industry.

Fifty Thousand Dollars

Edison's laboratory. Two large boxes roll onto stage. In one is Thomas Edison; in the other, Tesla. From within Edison's box comes the noise of his furious inventing.

Tesla's head emerges.

Tesla: I have an idea.

Edison: How's it coming there?

Tesla: I beg your—

Edison: What?

Tesla: I said, I—

Edison's head pops up.

Edison: I said, how's it coming?

Tesla: Surprisingly quickly.

Edison dives into his box, picking up his pace. A sudden crash from within.

Edison: Damnation!

Tesla: (*Suddenly*) Alternating current, Mr. Edison.

Edison: Eh?

Tesla: Alternating current—

Edison stops working.

Edison: Listen Tesla, we've . . .

He pulls Tesla closer.

Edison: Listen Tesla, we've gone through this before: Alternating current is not practical.

Tesla: But I've proven its efficiency. Surely a man of your intellect can see—

Edison: No, no, no—you're not listening! It's not practical. All my equipment is direct current; all my generators provide direct current! It's not about efficiency; it's about what people know and trust and what everybody else has. Who's going to buy an alternating current motor when all the power you can get is direct current?

Tesla: Provide alternating current. People will buy what works the best.

Edison: People will buy what everybody else is buying. And everybody is buying direct! If it ain't broke . . .

Edison pushes himself away from Tesla and goes back to work.

Tesla: *(To himself)* I assure you, it's broke—en.

Edison: What?

Tesla: Finished!

Edison: Finished?

Tesla: Finished.

Edison: You can't—let me see!

Edison climbs out of his box and barges into Tesla's.

Edison: Astounding! Look at them hum! Who would have ever thought to—my son, you deserve a raise.

Tesla: In addition to the fifty thousand dollars?

Edison: Eh?

Tesla: In addition to the fifty thousand dollars we agreed upon.

Edison: Oh, uh, ha, ha! Tesla, you misunderstood.

Tesla: Fifty thousand is a very clear figure.

Edison: No. No, I, I was . . . joking. American humour.

Edison climbs out of Tesla's box.

But seriously, I think a significant raise is in order.

Edison stamps a huge logo on the box that reads, "An Edison Original."

Tesla: (*Climbing out of his box*) It makes no sense to give a raise to the competition. Mr. Edison . . . I quit!

Edison: Oh . . . Is that a Serbian joke?

Tesla: Goodbye.

Edison: Don't be rash.

Tesla: Pizdo jedna! Pizdo jedna!*

Tesla exits. The boxes disappear.

Laboratory Spectacular

Narrator: Once a team; now a rivalry. Two inventors, both aware of the same conundrum: without application, technology cannot advance; without advancement, technology fails to be applied. Do you leave the drafting table for the marketplace in pursuit of capital? Or do you go the other direction, burrowing further and further into the remotest sanctum of your lab—blind to the world but able to spin your dreams in isolation, unsoiled by the dirt of men's hands?

Tesla's laboratory. Suspenseful music. Robert and Katherine speak in pitch blackness.

Robert: Mr. Tesla? Hello!

* Pronounced "Peezdo yedna" Translation: "You cunt!"

Katherine: Yoo hoo! Mr. Tesla?!

Robert: It's Robert Johnson, *Century Magazine.*

Katherine: This is ridiculous! He's not even here.

A neon sign lights up that reads: ABANDON HOPE ALL YE WHO ENTER HERE.

Robert: Oh, look at that!

Katherine: I think it's a sign.

Robert: Oh, pshaw! Let's go in.

Katherine: Mr. Tesla! Are you there? It's very strange to invite guests in with not so much as a candle—

Blazing white light. Tesla stands with electrodes in both hands as two million volts pass through his body. He screams, they scream. Blackout.

Katherine: Dear Lord!

Robert: Good God! Mr. Tesla! Are you all right?

Tesla: No.

Katherine: What?

Tesla: That effect should have lasted at least ten seconds.

The lights come up with a throbbing hum.

Tesla: Please, let me demonstrate again. Perhaps I'll produce a better display.

Katherine: Heavens, no!

Tesla: But this time I'll let your husband hold the electrodes.

Robert & Katherine: No!

Tesla: Very well. Welcome to my laboratory.

Robert: Mr. Tesla, may I introduce my wife, Katherine.

She holds out her hand. He doesn't take it but bows gracefully instead.

Tesla: Charmed.

Robert: How much current was running through your body just now?

Tesla: Two million volts.

Katherine: Through your body?!

Robert: Two million volts! You're pulling our leg!

Katherine: You should be dead!

Tesla: If I ran two million volts of direct current through my body, I would be dead. But this is my invention: alternating current. And it rushes back and forth at the rate of a thousand times a second. At this incredible frequency, the current just moves along the outer surface of my flesh—you could say, it doesn't have the time to go any deeper into my vital organs.

Robert & Katherine: Goodness! My, my!

Tesla: Also, the brain itself can only work so fast, and at frequencies above five hundred, it can't detect the stimulus, so there are no pain impulses sent to my brain.

Robert: Are you saying that *might* hurt, but you can't tell?

Tesla: Exactly!

Katherine: What a risk to take, to test such an idea!

Tesla: Testing implies failure. If I failed, I would be dead. So my experiments merely confirm what I already know to be true.

Katherine: You make it sound as if you were perfect.

Pause.

Tesla: Would you like to see the rest of my laboratory?

Shift. The trio stands at a long water tank.

Tesla: If you thought two million volts was shocking, you'd better brace yourself for this.

A remote control boat appears, operated by Tesla.

Robert: Interesting. A toy boat.

Tesla: Ha! It's not a plaything. Watch it on its course: now it will turn to the starboard, now port! Now starboard! Now port! Clever devil!

Katherine: It's incredible—how are you . . . ?

Robert: How can you . . . ? There's no connection!

Tesla: No. I call it teleautonomics. Control from a distance. Here Mrs. Johnson, try it.

He passes the controller to Katherine.

Tesla: This turns the boat to the right—this to the left.

Robert: I can't believe my eyes!

Katherine: Look at it go! You are quite something Mr.—

Katherine screams as the boat comes racing towards her. Tesla retrieves the controller in the nick of time.

Robert: I'm . . . I'm . . . Gadzooks, man! This is beyond me! How is this happening?

Tesla: I've discovered that alternating current at very high frequencies can produce an effect at a distance, without wires, if the receiving device is tuned to the same frequency.

Robert: Sensational!

Tesla: My next step is to apply it to communications.

Katherine: How do you mean?

Tesla: I'm going to transmit the human voice through the ether without the help of wires.

Robert: Now, just a minute! Words without wires? I can't believe that's possible.

Tesla: No, of course not. And one day you won't believe you couldn't.

Katherine: Do you think a person could receive information like this? I mean, thoughts. Wirelessly? From miles away?

Tesla: Thoughts? Why not? With the mind acting as a receiver, you mean? It's a fantastic idea! I can't wait to believe it. To discover the frequency of thoughts . . .

Katherine: My God! Mr. Tesla! I think I have!

Shift. Tesla holds an X-ray photograph and shows it to Robert and Katherine.

Tesla: Guess!

Robert: Strange.

Katherine: It looks like a head.

Tesla: Yes.

Robert: Is it a drawing? Or an ink blot?

Tesla: It's a photograph.

Robert: Of what?

Tesla: Of me.

Katherine: Oh dear, I'm afraid you're not very photogenic.

Robert: No, what is it?

Tesla: Me! My head. And this: this is my brain.

Robert: I can't comprehend it.

Katherine: Robert, it's a metaphor.

Tesla: No, it's a photograph of what my skin and bones are hiding. It changes how you perceive me, doesn't it?

Katherine: Yes it does.

Tesla: I become a bit of a machine, don't I? And just think, once this is widespread—oh God!—we'll know each other so much more intimately, no? I think we'll have a more truthful view of what it is to be human.

Robert: Can I have my photograph taken?

Tesla: Certainly. Just place your head in here and I'll turn it on.

Tesla directs Robert to an X-ray machine.

Robert: For how long?

Tesla: This exposure took forty minutes.

Robert: Forty minutes?! I guess it's worth it, for a picture of my brain.

Robert puts his head inside and Tesla turns it on.

Katherine: I hope you're not disappointed. Is it dangerous?

Tesla: No. Quite harmless—even beneficial. Afterwards, I feel my senses are calmed, and there is even a pleasantly detectable hum.

Shift. Tesla and Katherine are face to face. In the background, Robert remains with his head in the X-ray machine. Tesla writes numbers hidden from Katherine's view.

Katherine: Six?

Tesla: Yes. Again.

Katherine: Eighteen?

Tesla: Yes! And now.

Katherine: Three!

Tesla: Yes!

Katherine: *(Gasps)* Did I really guess right all those times?

Tesla: Of course!

Katherine: This is wonderful! It's as if, somehow we're joined. Our minds are fused!

Tesla: Like an electrical connection.

Katherine: Something more than just plain science.

Tesla: Plain science can explain a great deal—everything!

Tesla casually adjusts an electrical component and receives a violent shock. He is thrown to the floor. Katherine shrieks. Tesla suddenly revives, leaps to his feet and bows elegantly before the device.

Tesla: Electricity. She is always surprising me.

Katherine: You love it here, don't you?

Tesla: Yes. I'm married to my laboratory—or more specifically, to inventions—or *more* specifically, to electricity. She demands all of my passion.

Katherine: All your passion? You're a bachelor?

Tesla: Yes. And unlike an artist, a painter, or a writer like your Robert who may draw his inspiration from his love, the inventor who married would, by nature of his personality, give all his passion to his wife and be left unable to create.

Katherine: Oh, pshaw!

Tesla: I cannot think of many great inventions made by married men.

Shift. X-ray image of Robert's face. The hum from the machine rings in his ears, and the muffled conversation of Tesla and Katherine is in the background. As the sound grows, Robert's eyes widen in panic. He pulls his head out. Shift to stage.

Robert is physically charged as he stumbles from the X-ray machine.

Robert: *(Hallucinating)* Argh! Oh, shhh! Not so loud. Everyone else is still sleeping. Oh, they've fallen everywhere. Can you help me pick them up?

Robert gathers up invisible papers.

Katherine: Robert, what's happening?

Robert: Just gather them up and I'll alphabetize them all later.

Katherine: What?

Tesla: Would you like to sit down, Robert?

Robert: The poems. All the poems.

Katherine: What's happened to him?

Tesla: Perhaps he's experiencing a reaction to—

Robert: What are you two muttering about?

Katherine: Robert—

Robert is staring at his empty hands.

Robert: Oh, uh, I had a collection of poems right here.

Tesla: Yes?

Robert: I had been writing, I think, inside your machine, and the poems just materialized complete from start to finish. And I was just merrily transcribing away in there. And as each poem flowed effortlessly onto the page is was as though another was lined up patiently behind it. And the pages, I felt, were just falling out behind me, like a wake behind a boat!

Tesla steps forward and addresses the audience.

Tesla: This is precisely how I feel about my inventions. When I get the idea, I start at once building it up in my imagination. I change the construction, make improvements, and operate the device in my mind. It is absolutely immaterial to me whether I run the turbine in my thoughts or test it in my shop.

Scene resumes.

Robert: Oh! I've never felt such joy, such pure inspiration—Mr. Tesla, you've invented an inspiration machine!

Tesla: Well, the photograph is not quite ready yet. The emulsion plate must be treated in a chemical bath.

Robert: Forget the photograph; I just want the little box! I could have lived happily in there forever except that my leg cramped up on me.

Katherine: Oh Robert! That's wonderful! Can you remember any?

Robert: Oh I remember them all, and there were several dozen, including an epic with seven hundred and twenty-nine couplets with an unusual rhythm of alternating thirteen and fourteen beats per line.

Tesla: That's nineteen thousand, six hundred and eighty-three syllables, which is three to the power of three to the power of three. Very interesting.

Katherine: That's incredible! How did you—

Robert: Ah! I invented my own form of sonnet, with its own rhythm and rhyme scheme. My own sonnet! A Johnson. Katherine, I have a Johnson!

Katherine and Robert cheer. With a sudden start, Tesla checks his pocket watch.

Tesla: Robert, Katherine, please, don't think me rude, but I must bring this scintillating conversation to a close. You see, in precisely six minutes I am due to entertain a large crowd in Central Park.

Katherine: Oh? What's the occasion?

Tesla: Dinner. A feast!

Katherine: Dinner! In Central Park? With whom?

Tesla: Pigeons, hundreds of them.

Robert: Pigeons?

Tesla: I am close to each bird in different ways but there is one—she is very special—we spend hours together, laughing and . . . She is all white with grey tips on her wings, very beautiful, very passionate. These birds have extremely sophisticated methods of communication. *(To Katherine)* We speak of the mind as a receiver, for instance, well, these creatures are equipped with the most sensitive . . . But I shall keep them waiting.

Robert: Of course. We shall leave you.

Katherine: Mr. Tesla, I shall never forget this!

Tesla: Think nothing of it, my dear. Really, I must go. Thank you. Good evening.

Robert & Katherine: Good evening.

Robert and Katherine exit. Urgently, Tesla pulls out a bag of birdseed and meticulously rolls the bag open, folding it in patterns of three. He journeys to Central Park, and calls for his beloved Pigeons, who flock to him. He feeds them and speaks to himself, recalling the evening's conversation.

Tesla: Sensational! My next step is to apply it to communications. How do you mean? I'm going to transmit the human voice without wires. Words without wires? I can't believe that's possible! *(Pause)* Mr. Tesla, I shall never forget this.

Tesla then imagines how the press and public will perceive him.

Although he remained an enigma, as few could get near him, it only served to increase the public's interest.

The Pigeons flock like socialites at a high-society party.

Tesla the sorcerer, the alchemist, the high priest of science!

A sudden gesture frightens the flock away but for the White Pigeon. She lands at his feet, he removes his glove and strokes her affectionately.

A man inventing at a furious rate. Dozens of entirely new patents every month—modifications to existing inventions daily.

The Narrator joins Tesla on stage and takes over his speech.

Narrator: Other inventors became like pigeons at his feet picking up each unique patent and investing a career of thought into ideas that seems to spill from him like birdseed.

The Narrator makes a sudden move to frighten away the White Pigeon. He draws closer to Tesla, who begins to manifest the symptoms that the Narrator describes.

But what could they see at such a distance? Some peered closer, only to be further repelled by the complexity of the enigma. Signs of his eccentric personality began to surface. Talk of trepidation and anxieties. Germ phobias to rival Howard Hughes. A hatred of pierced ears, and an aversion to peaches. He stated that he would never touch human hair except perhaps at the point of a revolver. Any series of things he did had to be divisible by three and if interrupted, he would have to start again from the very beginning. He suffered from an acute sensory stimulation and he said dropping small squares of paper into a dish of water gave him an unpleasant taste in his mouth. No one could answer why he would even be doing that. More terrifying still, a published interview did not dispel these rumours, but cited them as the cause of his success! He had visions so vivid he often couldn't distinguish them from reality. He worked almost entirely in his head, building inventions and entire systems, setting them running, making improvements, perfecting, and finally, only when he was sure what he had created would work, would he draw the diagram. And once constructed, it always functioned exactly as predicted. He only had time for his work. And few could get near him, but if someone did . . .

The Muse

Robert and Katherine are revealed by the fire in their drawing room. Katherine holds a bouquet of roses.

Robert: Come, Katherine, have a seat by the fire. I'd like to share with you my latest creation. It begins: "In Tesla's Laboratory. The darkened figure atop electric spire. His hands ablaze with Promethean Fire."

This is the first of several classical allusions. I've also tried to emulate the sound of electricity as it rushes back and forth across the surface of the skin, with a cyclical repetition of words. It goes: "Oh—"

He becomes silent, but continues his recitation to Katherine. Her attention has shifted to thoughts of Tesla.

Katherine: Dear Nikola, I've tried several times to thank you for the roses, they are before me as I write—so strong, so superb in colour. We are wondering if anybody is coming tomorrow to cheer us up? We are very dull and very comfortable before an open fire but two is too small a number. For congeniality there must be three.

Katherine begins to levitate.

Sometimes I hope you will make me tell you what I know about thought transference, one would need to feel herself en rapport to speak of such things. I have had such wonderful experiences, I sometimes fear it will all pass away with me and you of all persons ought to know something of it for you could not fail to have a scientific interest in it. I have often wished and meant to speak to you of this but when I am with you I never say the things I had intended to say, I seem to be capable of only one thing.

Katherine floats back to her seat.

Do come tomorrow for my sake, as I need cheer from who is all-potent as you. Katherine.

Katherine directs her attention back to Robert and he is suddenly audible.

Robert: Yes, I think that's pretty good.

They smile at each other. Blackout.

Battle Of The Currents

Narrator: Nikola Tesla and Thomas Edison were vying for the contract to build the first-ever hydro project.

Image of Niagara Falls.

The award: a million dollars in royalties. The site for this historic event: Niagara Falls. But the honeymoon for this couple had been long over. It was AC versus DC in "The Battle of the Currents."

Fanfare. Spotlight. Thomas Edison stands before a noisy crowd with two Showmen at his side. He produces a Slinky.

Edison: Ladies and Gentlemen: we are in the presence of a monster. A terrifying aberration of nature: deadly, unpredictable, *alternating current.*

He passes the Slinky off to his Showmen who stretch it across the stage. At his signal, they manipulate it to create an oscillating waveform that spans the stage.

First it's going *this* way, then it's going *that* way . . . *this* way and *that* way . . . who knows which way! Who can predict? It's unpredictable. And deadly.

Edison produces a kitten.

Who doesn't love a kitten? Whose heart doesn't melt to touch its soft fur? I love a kitten; you love a kitten; we all love a kitten! Could anyone not love kittens, you ask. Well, alternating current doesn't love a kitten! Oh no. Cruel and vicious, alternating current hates little defenseless kittens.

Edison places the cat onto a large metal plate.

Nice kitty, good kitty. Here, Ladies and Gentlemen, is a Westinghouse AC generator. Feast your eyes! Witness pure evil incarnate! Attached to this generator is a metal plate. The kitten stands upon the metal plate. I turn the generator on as one thousand alternating volts rush through the—oh my Lord! . . . It's horrible . . .

The kitten cries, so Edison cranks up the voltage.

. . . Two thousand, three thousand, five thousand alternating volts rush through the kitten . . . Heavens! The humanity!

Edison shuts down the generator. He pokes at the kitten with his foot to make sure it is dead.

I warn you, though. What happened to this poor creature can happen to you! You too will be *Westinghoused,* if you're not careful. Help me fight against this merciless killer. Help me exterminate alternating current!

Tesla has appeared in the audience.

Tesla: Mr. Edison, wouldn't the same thing happen if it was a DC generator?

Edison: Eh? I'm sorry. I'm a little hard of hearing. Next question . . .

Tesla: I said, a DC generator might do the same thing!

Edison: Oh. Ha! No, madam. Never! Never! In the first place, I can't imagine anyone hooking a DC generator up to a kitten. But to give you a more direct answer—listen to this:

Edison cups his ear and we hear a faint Morse code signal. An old phonograph plays, Edison sings. He and his Showmen tap dance to the song.

Be Direct With Me Darling

Intro:

What's that humming from above?
The wires are buzzing with
Transmission of your love.
Between you and me,
There's an e-lec-tri-ci-ty.
I think I know your heart,
But I require some spec-i-fi-ci-ty . . .

Chorus:

Be direct with me, darling,
And I'll be direct with you.

Be direct with me, dearie,
And I'll be direct with you.

If you adore me,
Don't ignore me,
Just implore me
To say "I do! I do!
I do! I do! I do!"

And if you're direct with me, darling,
I'll be direct with you.

When you're alternating, vacillating,
How can I cope?
You shoot in all directions
And you scare me.
But when you're direct with me, darlin',
You fill me with hope.
Direct and to the point
Is how to snare me.

Dance break.

Your smile is a light bulb
I think it's the right bulb
For me.
I'm swimming in the current of your eyes
All the way to Washington, DC.

Edison opens his jacket to reveal the letters D and C emblazoned on the inside lining. Singing a reprise of the chorus, they begin to dance towards the wings.

Tesla storms in, interrupting the song.

Tesla: If this is direct current, allow me to show you what alternating current can do!

Tesla throws the switch of the generator flooding the stage in bright light. In a blaze of rhythmic steps, he seduces Edison's Showmen into a wild Serbian dance. Edison slinks off stage. Tesla finishes with a flourish.

Živeli!* Nikola Tesla and alternating current: Two names that America won't soon forget!

Cheers and applause.

With Tesla bowing in the spotlight, the Narrator appears.

Narrator: With the winning of the contract to Niagara Falls, Tesla was poised to become the first ever billionaire as every manufacturer in the world embraced Tesla's system as a fundamental first principle. But these royalties threatened to destroy the businessman who backed him. Mr. George Westinghouse.

The Narrator hands a contract to Tesla.

The agreement promised Tesla a dollar per kilowatt sold. Westinghouse was taking on water and was soon to sink with Tesla's success.

Tesla addresses Westinghouse.

* Pronounced "Zheevelli" Translation: "Cheers!"

Tesla: This assures my wealth forever. And you will be struggling to survive. I suppose that's how business works . . .

He rips the contract.

But I am not a businessman. Your belief in my ability is all the investment I need. To your health, George, and your enduring success!

Tesla exits.

Narrator: It was a grand gesture. It was a visionary's dream. And it ensured that Westinghouse would become the empire that it is today.

Now an endless stream of ideas was pulling Tesla in a multitude of directions, but each seemed too small, too narrow to demand all of his focus.

What if for each moment in history there exists the opportunity for one perfect invention that has the power to change the direction of everything? One invention. One thought. How do you single it out from all the others?

Interference

Katherine, Robert and Tesla are at dinner at the Waldorf. Tesla obsessively cleans his place setting with a stack of napkins presented by a Waiter. Throughout the scene there are shifts where the action on stage depicts what is happening inside Tesla's brain. In these, the trio performs a rhythmic choral piece consisting of numbers. Throughout dinner, Tesla progressively becomes consumed by his calculations.

Katherine: Brilliant! Brilliant! / Nikola, Bravo!

Robert: Genius! What a day! This day will go down in history!

Katherine: Did you see the look on the President's face? I've never seen anything so bright!

Robert: Do you know who couldn't tear herself away from your pavilion at the World's Fair? / Anne Morgan.

Katherine: J. P. Morgan's daughter. / She's in

Robert: J. P. Morgan's daughter.

Katherine: love. Madly in love with you. / A woman can tell

Robert: She's quite the catch.

Katherine: these things. / At least I can.

Robert: You could do worse.

Katherine: You certainly have your pick / of the crop.

Robert: Everyone loves you.

Katherine: Not more than us.

Robert: No, indeed. You're ours.

Katherine: All ours. Forget Anne Morgan—she's a tart. / Give us a

Robert: Tart. Good God.

Katherine: private—she is too! Give us a private show tonight!

Robert: Yes give us another charge!

Katherine: Let us back into your brain.

They toast. Shift. Inside Tesla's mind.

	*	*	*	*	*	*	*	*	*	*	*	*	*	*	*	*	*	*	*	*	*	*	*	*	*	*	*
Tesla																1			1			1			2		3
Robert	1	1	1	1	1	1	1	1	1	2		3		3		1			1			1			2		3
Katherine	1	1	1	1	1	1	1	1	1	2		3		3			2	2		2	2		2	2	2		3

	*	*	*	*	*	*	*	*	*	*	*	*	*	*	*	*	*	*	*	*	*	*	*	*	*	*	*
Tesla		3			2	2		2	2		2	2	2		3		3		1	2	2	1	2	2	1	2	2
Robert		3		1			1			1			2		3		3		1	2	2	1	2	2	1	2	2
Katherine		3			2	2		2	2		2	2	2		3		3		1	2	2	1	2	2	1	2	2

	*	*	*	*	*	*	*	*	*	*	*	*	*	*	*	*	*	*	*	*	*	*	*	*	*	*	*
Tesla	2		3		3				3			3							3			3					
Robert	2		3		3		1	2	3	1	2	3	1	2	1	2	1	2	3	1	2	3	1	2	1	2	1
Katherine	2		3		3		1	2	3	1	2	3	1	2	1	2	1	2	3	1	2	3	1	2	1	2	1

	*	*	*	*	*	*	*	*	*	*	*	*	*	*	*	*	*	*	*	*	*	*	*	*	*	*	*
Tesla		3			3					3	3	3	3	3	3	3	3	3			3			3			3
Robert	2	3	1	2	3	1	2	1	2	3	3	3	3	3	3	3	3	3			3			3			3
Katherine	2	3	1	2	3	1	2	1	2	3	3	3	3	3	3	3	3	3			3			3			3

Shift. Waldorf Hotel. Katherine is in mid-sentence.

Katherine: —and how about Mr. Ferris's wheel? What an absolute sensation!

Robert: Thursday, dinner?

Katherine: You can see all of Chicago from the top!

Robert: Can you speak at the fraternity Sunday?

Katherine: I was almost crushed!

Robert: Anton Dvořák wants to meet you.

Katherine: Robert held onto me like a squid.

Robert: I need your advice on a dispute with a—squid?

Waiter: More wine, sir?

Tesla: No. Thank you.

Robert: Well why don't you just leave it on the table.

Waiter puts the bottle down. Shift.

	* *
Tesla	fif teen thir ty six ty
Robert	
Katherine	
Waiter	4 3 2 1 2 3 4 3 2 1 2 3 4 3 2 1 2 3 4 3 2 1 2 3 4 3 2

	* *
Tesla	1 2 3 4 six ty 1 twen ty 1 eigh ty 2 for ty 3 hun dred 1
Robert	1 2 3 4 six ty 1 twen ty 1 eigh ty 2 for ty 3 hun dred 1
Katherine	1 2 3 4 six ty 1 twen ty 1 eigh ty 2 for ty 3 hun dred 1
Waiter	1 2 3 4 six ty 1 twen ty 1 eigh ty 2 for ty 3 hun dred 1

	* *
Tesla	six ty 1 twen ty 1 eigh ty 2 for ty 3 hun dred 2 six ty 1 twen ty 1 eigh ty
Robert	six ty 1 twen ty 1 eigh ty 2 for ty 3 hun dred 2 six ty 1 twen ty 1 eigh ty
Katherine	six ty 1 twen ty 1 eigh ty 2 for ty 3 hun dred 2 six ty 1 twen ty 1 eigh ty
Waiter	six ty 1 twen ty 1 eigh ty 2 for ty 3 hun dred 2 six ty 1 twen ty 1 eigh ty

	* * * * * * * * *
Tesla	2 for ty 3 hun dred 3
Robert	2 for ty 3 hun dred 3
Katherine	2 for ty 3 hun dred 3
Waiter	2 for ty 3 hun dred 3

Shift. Tesla experiences a startling visionary insight. Unable to speak, he rushes out, ecstatic. Robert and Katherine are dumbfounded.

The Meeting

Projected is a short film, with voice-over narration by Tesla as an old man.

Music. Tesla stands alone at night on an open desert plain taking in his new surroundings. Lightning flashes far off on the horizon.

Tesla: *(Voice-over)* It was in 1899, I think . . . Yes . . . 1899. I had a vision . . . a vision that I was no longer in the city, but in a wide-open place in the countryside. A lightning storm was flashing on the horizon . . . powerful bolts of electricity that seemed to be in some way . . . energizing the earth itself. And as I watched this in fascination, a man appeared to me.

The Narrator appears in the distance running toward Tesla.

He seemed to come out from the very centre of the storm . . . as though . . . the lightning brought him into being.

He has come close enough now that we see who he is.

Funny, he looked a little like Edison.

The Narrator has arrived, breathing hard, excited.

I asked him where we were.

The Narrator speaks and points toward where he just came from. His gestures are filled with excitement.

(Whispering) The future . . . I am from the future.

The Narrator speaks passionately to Tesla.

He wanted to thank me for all I had done. "I am your legacy," he said. "I am your greatest invention." And then he showed me a marvellous thing.

A shot of the Narrator's hands. A black iron rod lies across them. Suddenly he is on all fours, digging in the sand. As Tesla watches in fascination, the Narrator drives one end of the rod into the ground then stands beside it proudly gesticulating. Tesla is confused. In slow motion the Narrator reaches out and clasps the end of the rod. He begins to glow. Everything glows. It's as though electricity is travelling directly from the earth through the Narrator and into the night sky.

He grew brighter and brighter until I could no longer see.

Whiteout. Blackout.

Colorado Springs

In the darkness, the sound of a projector switching on is heard. It runs throughout the scene. Music plays—the accompanying soundtrack to a silent film.

The following scene is played as though it is three short scenes from a newsreel depicting Tesla's year of work in the desert at Colorado Springs. Choppy, sped up action takes place around a large Tesla coil, emulating the style of an early silent film.

Reel One: In the blackout, an intertitle card is projected centre stage: TESLA BUILDS A LABORATORY AT COLORADO SPRINGS.

Under Tesla's direction, two Lab Assistants make some final adjustments to the equipment at Colorado Springs. Financier J. P. Morgan looks on waiting for his cue to join in the unveiling of this new laboratory before the media.

Tesla and Morgan meet in front of the Tesla coil, flanked by the Assistants. Tesla waves off Morgan's offer of a handshake to celebrate the new lab. Flustered, Morgan quickly shakes the Assistants' hands and then they, in turn, unroll a ribbon for the official cutting ceremony. Morgan snips the ribbon, and everyone applauds.

Tesla waves the Assistants away to clear the frame for the photo opportunity with Morgan. The two men wave to the camera, chat, joke, and pose for the press.

Blackout.

Reel Two: Intertitle reads: BY JOVE! THAT TESLA'S A MODERN ZEUS!

Two Assistants scurry into view to examine the Tesla coil in preparation for experiments. Meanwhile Tesla has climbed up behind the Tesla coil and is visible above it.

A third Assistant enters and draws the attention of the others to the observing camera. Their dopey smiles and waves to "mom" and "America" are cut short by Tesla's demands that they get on with the electrical experiment.

Assistants One and Two examine the schematic diagrams while Assistant Three does some last minute adjustments to the electrical panel. Then the three assemble at the panel of throw switches and stand ready after a quick adjustment of the equipment.

Tesla commands that Assistant One throw the first switch and a shower of sparks begins to rain down from his Tesla coil, but it's not enough for him. He calls for the second switch and Assistant Two throws the lever, increasing the discharge. It's almost unbearable for all except Tesla, who is only just getting started. He calls for the third switch and Assistant Three obeys, turning the laboratory into a maelstrom of electricity. The Assistants use every ounce of their strength to hold their switches in the "on" position.

Suddenly a huge explosion blows the circuitry and sends the Three Assistants flying backwards, where they land sprawled on their backs. After a moment they get up in a mixture of fear and outrage.

Tesla thunders down to survey the damage and Assistants Two and Three storm out of the lab, frightened and angry. Tesla catches the remaining Assistant and sends him back to the control panel.

After Assistant One changes the burning hot fuses and Tesla has reclaimed his position above the coil, Tesla demands that the Assistant throw all three switches simultaneously. The Assistant tries to refuse but finally concedes to the inventor's demands.

The switches are thrown in rapid succession, and the place fills with sparks. The Assistant cowers in the corner while Tesla revels in his glory.

Blackout.

Reel Three: Intertitle reads: TESLA UNLOCKS THE SECRETS OF MOTHER EARTH.

Tesla, now alone in the lab except for the prying eye of the camera, enters from behind the coil absorbed in his work. While making adjustments to the machine, he notices that the cameras are still present and rolling. He attempts to politely address them while indicating he now needs to work.

In growing frustration he acquiesces to their demands that he pose by the coil in a "casual" stance for a bit of staged "candid" photography.

Finally, he can no longer bear the interruption and tells the crew that the interview is now over. When he discovers that they persist in shooting their newsreel, he becomes angrily protective of his work and tells them to take their equipment and be gone. Finally, he has no choice but to close off the lab from the paparazzi and when he bars the final door the film comes to an end leaving Tesla alone in his increasingly obsessive fixations.

Blackout.

Music. In a shadowy half-light, Tesla stands, where the previous scene left him, alone with the sphere and lost in thought. The Narrator appears from behind the sphere, holding, cupped in his hands, a glowing miniature version of the sphere, which will illuminate him during his speech.

Narrator: At the century's turn—with his inventions now transforming industry in America and around the world—Nikola Tesla was in the desert.

Tesla observes the vast desert plains and "conducts" the distant lightning storm with his hands.

His laboratory, closed to the public, stood alone on a vast plane. Residents in the nearby town of Colorado Springs frequently reported strange phenomena: random and oddly isolated lightning storms, entire fields becoming electrified and cattle electrocuted while grazing. Among these locals, Tesla inspired fear and awe.

Lightning strike and thunder clap. Music stops. Blackout. After a beat the Narrator's light returns.

Well, mostly fear.

Music returns.

But back home, rumour spread and anticipation grew.

Today what we know of his experiments in Colorado are contained in an obscure and cryptic set of notes. The diagrams are elusive, the math is, well, non-existent—but the findings . . . the findings . . .

Broadcast power! Electricity sent without wires. Harness the electrical potential of the earth's ionosphere . . . Send electricity through the earth itself . . . Unlimited, free, clean energy . . . Available anywhere!

Blackout.

Dinner Scene

From the darkness.

Tesla: Don't touch me!

Lights up. One year later. Tesla has just arrived at the Underwood-Johnson residence and stands in the doorway. The trio stands in an awkward moment, which Tesla breaks.

Good evening, good evening. *(Pause)* I beg you forgive my tardiness. This day has been most irregular.

Robert: Won't you come in?

Tesla: In?

Robert: Inside.

Tesla: Pardon me. I'm out of sorts.

Katherine: You look—if I may say so—very different, Nikki.

Tesla: I am different.

Robert: Well, dear friend, your arrival is all Katherine's talked about for weeks! Shall I take your coat?

Tesla: No! Thank you, Robert, no. Best to leave it.

Katherine: Here. Come this way. I've prepared all your favourite dishes.

They move into the dining room. Tesla is immediately drawn to the window.

Tesla: I have been walking.

Robert: Oh, yes.

Tesla: Listen to that. The city. The engines. It's growing so quickly. The lights in the city are beginning to hide the light of the stars. Did you know that? Three hundred variations of my patents I counted—now in general use about town!

Robert: That's wonderful.

Tesla: It's horrific! I had no idea things would progress so fast!

Katherine: Please, Nikola, calm yourself. Robert, dear, pour our guest some port.

Robert: Yes, and tell us about the desert.

Robert pours port for Tesla and Katherine.

Tesla: Yes, the desert. A whole year and what do I have to show for it? Nothing. No new patents, no improvements, nothing for sale. Also I have no money, and apparently I have been long forgotten.

Katherine: No! Not true!

Robert: Absolute rot! Pardon me a moment, won't you darling, while I produce some hard evidence.

Katherine: Excellent, dear!

Robert hands each a glass of port before scuttling off.

Katherine: It is madness for you to think this way! Everyone speaks of you and wonders what you're up to. They all ask me, because they think I know, but what can I tell them? I say, "His correspondence was all pleasantries, and spoke little of his work"—a bare-faced lie, Mr. Tesla, because you never once responded to my letters!

Tesla: How I wanted to! But I simply could not.

Katherine: It was difficult enough just sending them.

Tesla: What do you mean?

Katherine: Well, you see, when I write I must always make several attempts—a system of repression—because I can never express what I would say. I hope you weren't offended when I wrote you last. I did not mean to be severe. I was only wrapped up in disappointment.

Tesla: You had every reason.

Katherine: I missed you very much, and wondered if it is always to go on this way. Can I ever become accustomed to not seeing you?

Tesla: Katherine, I . . .

Robert returns with a stack of magazines and newspapers.

Robert: Here we go! *The Citizen*, "Where is Nikola Tesla?" *The New York Times*, "Deserted Tracks Along The Tesla Trail." *Out and About*, "Tesla." A different magazine every month. We collected them for you.

Tesla: Very kind.

Katherine: So you see, we have single-handedly created the "In Absentia Society" for our dear absentee.

Tesla: Again, Robert, Katherine, you outdo yourselves on my behalf.

They toast. With his handkerchief, Tesla wipes his glass where it has touched the others.

You see, there is a great portion of me that is still working away while you see me sitting here before you. I confess I am totally consumed. Continuous, tortured concentration. At times I am the happiest of mortals and at times I am like a man condemned to death as a result. Yes, yes, truly I fear for my safety, Robert. The possibility of a blood clot or atrophy of the brain is a constant threat to me.

Unnoticed by Katherine and Robert, the sphere rolls through the front door and into the dining room where it hovers, distracting Tesla.

Ideas! Every day, struggling to drive out all the old images, which were like corks on the water, bobbing up after each submersion. But after weeks, months of this desperate cerebration I have finally succeeded in filling my head chock-full with my new subject.

Tesla is now totally fixated on the sphere.

Katherine: Subject?

Tesla: My greatest creation. It will make my earlier discoveries look like deficiencies.

Katherine: We had no idea.

Robert: This . . . this is extraordinary.

Robert pulls out a notepad.

You don't mind if I take notes.

Tesla: Notes on what?

Robert: Your description of this . . . whatever it is. I want to be the first / to proclaim . . .

Tesla: No! No! Robert! You misunderstood! The utmost secrecy must be shown in this regard.

Robert: Perhaps a statement, then.

Katherine: Yes, a message from the man who can control the heavens!

Tesla: What do you mean?

Katherine: Well, that's the reigning theory.

Tesla: Have you heard something?

Robert: Nothing but rumours.

The sphere disappears.

Tesla: What rumours?

Katherine: I read you blacked out Colorado Springs.

Tesla: You *read* this? Where?

Katherine: Why, in the *Century*, of course!

Tesla turns on Robert who is proudly holding up a copy of Century Magazine.

Tesla: Robert! You published this . . .

Tesla snatches the magazine from Robert and pores over it.

Tesla: Ah! Not . . . ! Oh! Indecent!

Robert: There isn't a problem, is there? Katherine said / you had given—

Katherine: I wrote you and asked your permission, you silly sot! Don't you recall? You didn't respond, so I—

Tesla: Of course! Of course! How could I . . . I could never forget! They made me so happy your letters. Yes, such a beautiful gesture, I always meant . . . It was never my intention . . . The truth is I never read them at all.

Katherine: Oh!

Katherine is visibly shocked but struggles to remain composed.

Oh. I see.

Tesla: No. No.

Katherine: What about the packages?

Tesla: Yes. They looked . . . marvellous.

Katherine: But you didn't open them.

Tesla: My dear I—

Katherine: No, no. Please sit down. No. Look at you! You're so embarrassed. Sit down.

The trio regroups at the dinner table.

Robert: What packages?

Katherine: Oh, little scraps of nostalgia, and so on. So he wouldn't forget us.

Robert: I didn't see these scraps.

Katherine: Well, and notes. Little things I scribbled quickly. Ideas and such. I always brought up the content in our conversations, anyway, darling.

Robert: Ideas.

Katherine: Yes.

Robert: How many of these "packages" did you send?

Katherine: I don't know. I didn't count them, Robert.

Tesla: Three a week. Always three. Very clever.

Katherine: You must not think me very clever if you don't open my letters.

Robert: Three a week?

An awkward silence. Katherine makes a sudden realization.

Katherine: Marconi!

Robert: What?

Tesla: What about Marconi?

Katherine: *(To Robert)* He won't have heard about Marconi.

Robert: You haven't heard about Marconi?

Tesla: I know him well. He's / worked with me.

Katherine: It was in one of my letters.

Tesla: What about Marconi?

Robert: He just sent the first wireless. He's become famous. And he was using seventeen of your patents.

Tesla is overcome by panic. He collapses to the floor in a state of seizure.

Nick?

Katherine: Oh God! Nikola! Fan him, Robert!

Robert: I'm fanning!

Robert fans Tesla. Katherine gathers Tesla's coat and hat, which have been scattered during his fit.

Tesla: No!

Katherine: Nikki! Are you all right?

Tesla: No! No! No!

Katherine: Nikki!

Tesla: Morgan was expecting / me to have completed . . .

Robert: Easy, easy.

Tesla: and I need more money, and now it's too late. / Wardenclyffe! Wardenclyffe is lost.

Katherine: What's he talking about?

Robert: I don't know. Nikki, Nikki! Be calm. Marconi's transmission was nothing.

Tesla: But he succeeded!

Robert: He only transmitted across the English Channel.

Tesla: The English Channel?

Katherine: And it was just / the letter—

Tesla: How many frequencies did he demonstrate?

Robert: I don't think / he even understands the concept.

Katherine: None.

Tesla: There must be at least one.

Katherine: Or one, I mean. And all he sent was / a letter.

Robert: He only sent a letter.

Tesla: A letter? To whom?

Katherine: No. Just a letter. / The letter *S*!

Robert: *S*!

Tesla: *S?*

Robert & Katherine: Yes!

Robert: In Morse code!

Tesla: Morse . . . not by voice?

Katherine: No.

Tesla: *S?*

Robert & Katherine: Yes.

Tesla: That's just dot, dot, dot.

Robert & Katherine: Yes!

Tesla: Ppt. Ppt. Ppt.

Robert & Katherine: Ppt. Ppt. Ppt.

Tesla is beginning to regain his composure.

Tesla: Perhaps I'm still safe then. Perhaps there's a chance.

Katherine: Of course! You are the only man. The only man who could do it.

Katherine helps him back into his coat.

Tesla: There is no time to dally. It will happen. You must see it.

Robert: Which?

Tesla: My secret. "In one of which a sumptuous temple stands, that threats the stars with her aspiring top; thus hitherto has Faustus spent his time."

Work must commence tonight! Will you join me on Long Island?

Robert: Long Island?

Katherine: Yes!

Robert: No!

Tesla: Come, let's fly!

Robert: What about dinner?

Katherine: I'll pack it. We can make it a picnic!

Robert: A picnic! Have you both lost your senses? You can't get a cab to go out there this time of night.

Katherine: He's right.

Tesla: How many bicycles do you own?

Katherine: Two.

Tesla: I noticed your neighbours own one as well.

Katherine: Oh, Nikola!

Robert: Well, I say! This really is the limit!

Tesla: Are you afraid, Robert?

Robert: Afraid! Of course I'm not afraid. It's just that . . . well . . . it's rather dark. And wet . . .

Tesla jumps up onto a dining room chair.

Tesla: "With swelling rains, foaming from rock to rock,
Along its course of ruin,
On to the inevitable precipice!"

Tesla exits.

Katherine: He's back.

Katherine blows out the candles.

Katherine: Come along Robert!

Katherine exits.

Robert: Abandon hope all ye who enter here.

Robert exits.

HAARP

The Narrator enters holding a dossier marked "Classified." He picks up one of the magazines featuring Tesla on the cover. He shows it to the audience.

Narrator: One of the characteristics of technology is that it always gets away from its creator and begins to live a life of its own.

A series of images appear depicting the HAARP experiments being conducted by the US military in Alaska. The Narrator references the top-secret dossier.

Jump ahead a century, same country, same technology, in a strangely similar isolated laboratory, an elite team of scientists working for the US Air Force and Navy begin work on a top secret project. Known as the High-frequency Active Auroral Research Project, or HAARP, their plan is to build a device that will blast so much radio energy at a specific portion of the earth's ionosphere that it will bulge into space. Their stated purpose: to increase communication efficiency between naval bases and their deep-sea nuclear submarines. However, the patents they're using outline techniques for experiments in weather manipulation, communication blackouts, and thought control.

The primary source for the patents . . .

The Narrator holds up the picture of Tesla.

Dr. Nikola Tesla, based on his experiments in Colorado Springs. One of the characteristics of technology is that it always gets away from its creator and begins to live a life of its own.

The Narrator covers Tesla's picture with the classified document.

Bicycles

Tesla, Katherine, and Robert enter on bicycles en route to Wardenclyffe. Tesla leads the way, Katherine follows enthusiastically, and Robert wobbles precariously on the machine.

Tesla: Invigorating, isn't it?

Robert: I've never felt quite comfortable on a bicycle.

Katherine: Look at me!

She throws her hands into the air.

Robert: Katherine! *(Back to Tesla)* The whole thing just doesn't seem likely. Balancing in this way. Two thin tires, no support—I feel it's only faith holding me up, and one day it's all going to topple over.

Katherine: What?!

Tesla: Don't lose faith!

Robert: What?!

Tesla: Faith! Don't lose it!

Katherine: What have you faith in, Nikki?

Tesla: I don't need faith. I have knowledge.

Robert: Look out!

Robert hits a divot in the road and nearly crashes. They ride on into the night. The lights fade as Tesla and Katherine pull ahead of Robert who is unable to keep up the pace.

The Tower

Darkness. Fog. At the foot of Wardenclyffe, a voice calls out.

Robert: Are you there?

Tesla: This way!

Robert: Thank God, I thought I lost you.

Katherine: Robert. Look.

Robert: Good God Almighty!

Katherine: What is it?

Tesla: Katherine, Robert, I give you Wardenclyffe.

Tesla flips a switch and lights suddenly come blazing on through the fog, illuminating the tower. Tesla's sphere is perched on top forming the shape of the legendary broadcast tower.

It is the reason we are here. It is the culmination of my work at Colorado Springs. It is Archimedes' proverbial lever! And it changes everything.

Robert: Changes?

Tesla: Everything. From this little point on Earth I will abolish the concept of space. From this moment I will banish the notion of time! And I have the two of you to thank!

Robert: Us?

Tesla: It's not quite complete yet, so I didn't want to mention it at first. I built it in absentia. I sent plans and diagrams to my assistants in New York and they built it. In faith.

Katherine: Oh, Nikki, I want to climb it!

Tesla: Splendid! What Katherine desires, Katherine must have. Come in, there's a ladder running up the inside of the tower.

Katherine: No. I want to climb it from the outside!

Tesla: The outside?

Robert: Have you gone mad?

Katherine: I must.

Tesla: Well . . . The structure does lend itself to easy ascension.

Katherine: Hooray!

Robert: Traitor!

Tesla: What Katherine desires . . .

Katherine gazes up at the tower preparing to ascend.

Katherine: Dr. Tesla, tell us!

Tesla: What?

Katherine: Tell us where we're going.

Robert: We're following you!

Tesla: (*Laughing)* Indeed, Katherine, it's time for you to tell us where we are going.

Katherine begins to lead the two men in climbing up the tower's exterior.

Katherine: We are going up. And the higher we climb the further we shall see. And look at us! Connected by this structure. Thoughts sent from Colorado to New York City now hold us in place, and should this whole thing fall . . .

Robert: Oh, Katherine, don't say it!

Katherine: We will have seen the future!

They climb as lights fade. The wind swells.

Lights reveal Tesla, Katherine, and Robert now clinging to the tower near its pinnacle. Robert is laughing, giddy with the heights.

Robert: Look at us! The strangest things happen when we're with you Nikki. Hanging a hundred feet in the air. I'm drunk, but we haven't opened the wine, and, oh my lord, look at the sky!

Katherine: It's coming, it's coming!

Robert: Oh, it will be sunrise any minute now.

Tesla: The future.

Silence. They are absorbed by the beauty of the coming dawn.

Robert: Nikki.

Tesla: Robert.

Katherine calls out the shifting colour pallet on the horizon.

Katherine: Purple.

Robert: What did you mean when you said this "changes everything"?

Tesla: While in the desert, I discovered I was wrong.

Robert: About what?

Katherine: Red.

Tesla: My polyphase alternating current system.

Robert: What about it?

Tesla: It's flawed.

Robert: Nikki, it's perfect. My God, you know that.

Tesla: No, no, Robert, it only appeared so when held against the monstrosity of direct current. But I've gained such perspective now. As if gazing upon my inventions from an enormous height.

Katherine: Pink.

Tesla: Now I have envisioned an entirely new system, one so great it required the planet as its laboratory and Wardenclyffe is its cornerstone.

Robert: A new system?

Tesla: A new system.

Katherine: Gold!

Robert: But, Nikola, people are hardly used to your old system; only beginning to comprehend its size.

Tesla: They're wasting their time. They're speculating the merits of a relic.

Robert: You know, even Edison is working with your motors and generators now.

Tesla: He's too late. I hereby sentence alternating current to the museum!

Robert: I don't know how easy that will be.

The growing sunlight intensifies.

Katherine: Here it comes! Look, look! There! There!

The sun appears over the horizon, illuminating their faces.

It's here! Can you feel us spinning towards it? I feel like I'm falling! Falling into the day. Falling into the sun!

Katherine reaches towards the sun. Fade to black. Sound of ocean waves. Lights up on the Narrator at the seashore. Tesla can be seen in the distance, atop his tower.

Narrator: What I find so compelling is that all across the ocean, the endless ocean, little waves are created everywhere in every direction by wind, boats, fish, rain—an infinite collection of unrelated acts. And they all interact in a multitude of ways: some enhancing each other, others cancelling each other out. But at the end, here at the distant shore, this is what we receive: a coherent pattern, a steady rhythm emerges, a pulse that draws order from disorder and defies that hateful second law of thermodynamics, which says everything must break down, cool off and die. Defies it and gives us knowledge of the inevitability of creation—gives us hope.

Reporters

Three Reporters have laid siege to Tesla's tower. They stand below assailing Tesla with questions.

Reporter 1: Mr. Tesla, if J. P. Morgan pulls out of Wardenclyffe, what do you intend to do?

Reporter 2: Is it true you've already spent one hundred and fifty thousand?

Reporter 1: That's one expensive pickle factory!

Laughter.

Reporter 2: Mr. Tesla, is it true that the deed to Wardenclyffe has been used as collateral for your room and board at the Waldorf Hotel?

Reporter 1: Where is your money coming from without Westinghouse?

Reporter 3: From Mars?

Laughter.

Reporter 1: What about your lawsuit with Colorado Springs?

Reporter 2: Are you still a taxpayer, Mr. Tesla?

Reporter 3: Well who needs taxes when you've got free power?!

Laughter.

The First Reporter rushes toward the audience with a spinning front page. The headline reads: WARDENCLYFFE: TOWER OF BABBLE? The Second Reporter follows, with the headline: ONE MILLION VOLTS BUT NO MILLIONAIRES. The Third Reporter concludes with: TESLA UNCOILED.

Narrator: Of all of Tesla's inventions, his unwitting contributions to the rise of tabloid media may have been the most enduring.

His failing finances had been lampooned in the press. Not to mention his belief that an advanced civilization on Mars was signalling him in New York.

Investors were withdrawing and his isolationist approach to business had left him without resources, backers, or partners. And with the competition closing in, time was running out.

There was one remaining avenue of support available. Financier and unabashed capitalist J. P. Morgan.

The Morgan Pitch

Inside the offices of J. P. Morgan, Tesla sits on top of the giant sphere addressing the famous financier.

Tesla: It stands in perfect isolation, towering above the Earth. At its crest of one hundred and eighty-seven feet rests the giant copper electrode, the voice box of the future.

I know you have heard of Marconi's helpless *S*. Well, if this is what you want, I will transmit the entire alphabet, every word it can spell, all at once, in all directions. Uniting the ears of humanity! Suddenly all distance is irrelevant. This will happen very soon. But I am not interested in making these little tests sending letters. Wasting my time with tiny steps when I stand on the brink of something far greater.

Swept away by his vision, Tesla deftly manipulates the sphere.

I live, Mr. Morgan, to see the day when electricity exists all around us like air and a man need only reach out and drink it in to his home. I live for the look on their faces as they stumble away from their ticker tapes and phonographs, from their meagre lamps and light bulbs, to feel the radiance of my beams washing over them. When voices pass between us like a conference of spirits. And when news of the world circulates like gossip at the office, and music plays without instruments in every corner. When pictures flash like dreams in the ether, and when the upper atmosphere is harnessed like a giant and the Earth's great power pulls us out of our timid lives and into the era of Titans!

Tesla stands with the sphere aloft: Atlas with his globe.

Yes, Titans, Mr. Morgan.

He rolls the sphere aside.

You could consider this to be an exclusive offer. There is an opening for an historical figure in Nikola Tesla's future. Join me and . . .

Morgan is clearly unimpressed.

Perhaps you find it surprising that a Serbian man has become America's greatest champion. Perhaps you're humiliated by this prospect. Or perhaps you simply have not the intelligence or perception to see what I am offering. Your vision is so clouded by your greedy pursuits of capital. I speak of a wealth for humanity, not that of a gluttonous . . .

Tesla loses his composure.

Svinjo jedna! Svinjo jedna! U pickŭ materina! Jebi se! Jebi se!*

He turns to leave. The sphere reappears, blocking his path. He turns to Morgan, desperate.

Tesla: Wardenclyffe is my flesh and bone, Mr. Morgan. And the construction is unbearably close to completion. If I can begin making convincing demonstrations I could . . . I am not a dreamer, but a practical man of great intuition, and my experiences are gained in long and bitter trials. The upper atmosphere harnessed like a giant and the earth's great power pulls us out of our timid lives and . . . If you join me they will remember you forever!

Suddenly the sphere bears down on Tesla. He senses it coming, looks over his shoulder and . . . Blackout.

Rockaway Beach

Sound of waves. Tesla alone at the seaside, lost in thought. Katherine and Robert enter, relieved to find Tesla.

Robert: I say! How about a dip? Tide's in. What's say we take to some bracing waters? We'll all feel better. Hey? Come on!

Tesla: You know, I almost think I could.

Robert: That's the spirit!

Katherine: Go along, Robert. We'll follow shortly.

Robert heads into the water.

Katherine: We heard about Morgan.

Tesla: Yes and . . .

* Pronounced "Sveeno yedna! Sveeno yedna! U peechku materina! Yebi se! yebi se!" Translation: "You swine! You swine! Go back to your mother's cunt! Fuck you! Fuck you!"

Katherine: I was so frightened. Today I went to the lab and the hotel. I asked for you everywhere. No one knew where you were.

Tesla: So how did you find me?

Katherine: When I finally stopped I felt the waves.

Pause. He looks at her then turns away.

Katherine: You don't need anyone, do you? How strange it is that we can't do without you.

Tesla: Look at him out there. Such a rare, vigorous gentleman.

Katherine: Oh, Nikola. I'm so sorry.

Tesla: Think nothing of it my dear, really.

Katherine: No. Yesterday he . . . he removed his name from your list of patrons.

Tesla: He did what?!

Katherine: He pulled out our shares. Some. The others are listed under a pseudonym.

Tesla: He did this?

Katherine: He did. And he encouraged his friends to do the same.

Tesla: But . . . how could he? Not Robert.

Katherine: Yes. And by betraying you he's betrayed me.

Tesla: The coward! Does he put that much faith in the bastards on Wall Street?

Katherine: Apparently.

Tesla: I see now. This is my fight and mine alone.

Katherine: I can fight! We can fight together.

Tesla: I hear them laughing at me. How it sickens me. J. P. Morgan thinks he has me licked. Well, I'll none of it! I'll create new worlds solely to punish him! Nothing can stop me!

Katherine: Of course! And I will support—

Robert returns, dripping wet.

Robert: Brrr! I suppose you were just going to let me shrivel up in there, were you? Where's my robe?

Tesla: Where's your backbone? That's the question.

Robert: What?

Tesla: Your spine!

Robert: What have you been saying, Katherine?

Tesla: First Morgan insults me with his callous incredulity, his obstinate, self-serving doubt, and now I must suffer this . . . this sabotage . . . from a friend.

Robert: A friend, yes. I am also the editor of a popular magazine.

Tesla: And I am the reason it is popular!

Robert: Agreed. You are a reason, yes. But, still, I must maintain some semblance of . . .

Tesla: Credibility.

Robert: Integrity. In this morning's *Times* you recount your latest theory . . . to reduce all human energy down to a scientific formula . . . it all seems so . . .

Tesla: So . . . ?

Robert: Well, quite frankly, it seems like something Katherine would dream up.

Katherine: Ha!

Robert: And there's nothing to that, you understand.

Tesla: Except that you can't publish it. Too threatening.

Robert: I beg your pardon?

Tesla: Then it is our friendship that is standing in the way of your work.

Robert: It's about readership, about perception.

Robert: People want to believe what you're saying, but they won't believe it if you can't support it.

Tesla: Of course they can't, and naturally I will!

Robert: I'm speaking as an editor now . . . Yesterday, as we sat at the luncheon table—some writers and myself—a young reporter / brought in an—

Katherine: No more. Please. For my sake . . . / stop it. Robert!

Robert: You have yourself to blame for setting this in motion.

Katherine: I was protecting him. You want to destroy him.

Robert: (*To Tesla)* A young reporter brought in an article where you predict an end to war in our lifetime. In the future, you say, war will be fought with robots shooting death rays. And I watched as the listener's faces turned first to curious smiles, then to smirks, then to disdainful laughter. My very own staff—all aware of my high regard for your person, and my friendship with you! To laugh in *my* presence! I cannot watch, as your name becomes a source of derisive entertainment where once it provoked fear and awe.

Tesla: You fear a scandal.

Robert: I fear nothing! I fear it's the beginning of something that will harm you greatly . . . unless you curb your dalliances with the press.

Tesla lunges at Robert, grabbing and shaking him.

Tesla: Send your assassins to the tower! They'll come back humbled by a force that is, quite literally, well beyond them. No one shall look at Wardenclyffe who is not at once jolted by a hundred million possibilities!

He releases Robert and moves away.

Robert: In my heart, I wish it could happen, but I / no longer believe—

Tesla: It is simply a matter of time for opinion to swing back. / You can't be concerned.

Robert: Time? How much longer / must we wait?

Tesla: Have you forgotten who I am and what got me here in the first place!

Robert: It's all too unpredictable! I can't risk it, Nick!

Pause.

Katherine: Look at him, Robert. What do you see? Perfection. I know how it eats at you. How you writhe at the thought that you will never possess it.

Robert: Ah, I see. Then why should he care if I turn away?

Katherine: Because he loves you.

Robert: Loves! He loves nothing but his own brain, into which you and I barely enter.

Katherine: Not true. / He just . . .

Robert: Some love! A man who doesn't even open your letters. Your precious packages.

Katherine: You're so transparent, Robert. So threatened.

Robert: Do you think I care about these past, forgotten glories. The Chicago Fair? When was that? Centuries ago! The Waldorf? The White House? Positively antediluvian!

Katherine: I remember.

Robert: Yes, you have to! These things are dead without you, Katherine. Outdated. Gone.

Katherine: Much like your poems.

Pause.

Robert: When's the last time you read one of my poems?

Pause.

Tesla: This is what comes of idleness. Casting doubts. I wish to formally end this conversation.

Katherine: Understandably. I never meant / for you . . .

Tesla: I know, I know. Now I must take my leave.

Katherine: No!

Tesla: Robert, I sincerely thank you for your honesty. My apologies, / Katherine, but . . .

Katherine: We have more to discuss!

Tesla: The time for discussions is over. I must get to work / as my work is paramount.

Katherine: What about our friendship?

Tesla: Valuable time is slipping away.

Katherine: You are not a machine!

Tesla: I am a machine! I must become a machine. I beg you both to pardon my suddenness.

Tesla leaves them.

Robert: Come, Katherine, don't interfere in this man's destiny. He's not like us mere mortals.

Robert places his hand on her shoulder.

Katherine: Don't touch me!

Blackout.

Disappearance

Inside Tesla's apartment, the exhausted inventor contemplates the sphere. The Narrator observes him for a moment, then gently pulls the sphere away. The sounds of pigeons suddenly fill the room.

Tesla: I have been feeding pigeons, thousands of them, for years. But there was one who was unlike all the rest. All white with grey tips on her wings, I would know her anywhere, I had only to call and she would come.

The White Pigeon flies in and alights beside him.

I loved her like a man loves a woman, and she loved me.

As Tesla touches her, the Pigeon becomes distinctly woman-like. Waltz music. They dance. Tesla's past glories rush by them until they are dancing in the hills of Budapest. As the music fades, the Pigeon transforms again and perches beside him, dying.

Narrator: "I had always known I would finish my life's work, no matter how difficult. But when that bird died something went out of my life."

The Pigeon flies off, the images of Budapest disappear. Tesla is back in his room. He turns to the Narrator.

Tesla: Bring her back.

Narrator: I can't.

Tesla: Why not?

Narrator: She's dead.

Tesla: Then begin again.

Narrator: I can't.

Tesla attempts to summon the past.

Tesla: "It's 1882 in Budapest, a young man is walking, his thoughts drift to a favourite passage from Faust and he speaks it aloud . . ."

Though Tesla continues to speak, he cannot be heard. The sphere appears above him, bearing an image of Wardenclyffe. As it lowers, Tesla is gradually obscured until he cannot be seen at all.

Last Letter

As the sphere rolls off stage, Tesla is revealed as an old man carrying a bundle of letters. It's 1943. He opens a letter and reads. Katherine appears.

Katherine: Dear Nikola, I came here a month ago, quite alone, to this hotel full, but empty for me. Here I am detached as if nothing belonged to me but memory. At times I am filled with sadness and long for that which is not—just as intensely as I did when a young girl and I listened to the waves of the sea, which is still unknown, and still beating about me. And you?

Tesla sees her before him and drops the letter.

What are you doing? I wish I could have news of you my ever dear and ever silent friend, be it good or bad. I do not know why I am so sad, but I feel as if everything in life had slipped from me. I think I would be happier if I knew something about you. You, who are unconscious of everything but your work and have no human needs.

This is not what I want to say and so I am faithfully yours, Katherine.

She fades.

Demolition

Narrator: July 4. American Independence Day. In an undeveloped district on Long Island known as Wardenclyffe, the sound of an explosion is lost amidst the fireworks and marching bands. Tesla's tower is destroyed by government dynamite.

Nikola Tesla's Tower at Wardenclyffe was destroyed as a wartime security measure. What dangerous things did they suspect he had discovered? Or was it that his tower was a monument to his ideas: free energy, a worldwide communication network for the people. Ideas that were best forgotten, ideas that shouldn't have a monument.

Explosion. Tesla and the Narrator watch as Wardenclyffe collapses. Blackout.

The End

Be Direct With Me
Words by Electric Company
Music by David Hudgins
Arrangement by Bill Costin
Slow Rubato Intro:
What's that hum - ming from a bove? The wires are buz - zing with trans
mis - sion of your love_ be-tween you and me there's an e - lec - tri - ci - ty I
think - I know your heart, but I re- quire some spe ci- fi - ci - ty
Ragtime feel:
Be di
rect with me dar - ling and I'll be di rect with you
Be di
rect with me dear - y and I'll be direct with you
If you a
dore me_ don't ig-nore me_ just im plore me to say I do. I do I

2
Be Direct With Me
do I do I do and if you're di-rect with me dar-ling I"ll be di-rect with you
When you're al-ter-na ting, va-se-la-ting how can I cope? You
shoot in all di-rec-tions and you scare me. but when you're di-rect with me dar-ling, you
fill me with hope. Di-rect and to the point is how to snare me... (Dance Break!)
Your smile is a

Be Direct With Me
3
light bulb I think it's the right bulb for me I'm
swim - ming in the cur - rent of your eyes all the way to Wash - ing - ton D.
Ritardando
a tempo
C. Be di - rect with me dar ling and I'll be di rect with you
Be di-rect with me deary and I'll be di rect with you
If you a-dore me don't ig - nore me just im - plore me to say I

4
Be Direct With Me
66
A♭m7 D♭7 C♭6 A♭m7
do. I do I do I do I do and if you're di-rect with me dar-ling
69
D♭m7 G♭7 C♭6 A♭7
I"ll be di-rect with you and if you're di-
72
C♭6 A♭m7 D♭m7 G♭7 C♭6 G♭7 C♭6
rect with me dar ling I"ll be di-rect with you

Electric Company Theatre is Kim Collier, David Hudgins, Kevin Kerr and Jonathon Young. A unique and vital presence in Vancouver theatre, the company strives to create life-affirming, inspiring and provocative work.

The members of the Electric Company Theatre met while training at Vancouver's acclaimed Studio 58 acting school. They have been creating their original brand of physically based theatre since 1992. Collectively their talents and skills include writing, direction, design, education and arts administration.

The company writes and produces original works of theatre with an emphasis on collective creation. Through this process they aim to create a rewarding working environment for artists, where artists are encouraged to think outside their traditional roles. Their plays capitalize on the immediacy, imagination and magic of live theatre.

Other plays created and produced by Electric Company:

The Palace Grand by Jonathon Young (Vancouver East Cultural Centre, 2004)
The Fall (Warehouse on the Finning Lands, 2003)
The One That Got Away by Kendra Fanconi (Swimming Pools—Jewish Community Centre, 2002)
Flop (Vancouver East Cultural Centre, 2002)
Dona Flor and Her Two Husbands with Carmen Aguirre (Vancouver East Cultural Centre, 2001)
The Score (Waterfront Theatre, 2000)
The Wake (Granville Island, Vancouver, 1999)
MUD (*Theatre Under the Gun*, Vancouver East Cultural Centre, 1999)
The Mammoth (*Theatre Under the Gun*, Vancouver East Cultural Centre, 1998)
Great Day for Up by Jonathon Young (More Four Play Festival, Studio 58 Theatre School, 1997)
Molly Brolly and the Folly of Love by Kim Collier (Women in View Festival, Vancouver, 1995)

Awards:

Nominated for fifty-five Jessie Richardson Awards and winner of twenty-seven awards for Excellence in Theatre including: four for Best Production (*Brilliant!*, *The Score*, *The One That Got Away*, and *The Fall*); two for Best Original Script (*Brilliant!* and *The Score*); Outstanding Performance by a Lead Actor (Jonathon Young, *The Palace Grand*); two for Best Direction (Kim Collier, *The Score* and *The One That Got Away*); Outstanding Emerging Director (2000, Kim Collier).

Winner of the Alcan Performing Arts Award 2001 (*Dona Flor and Her Two Husbands*).

Kevin Kerr won the 2002 Governor General's Literary Award for Drama for *Unity (1918)* (Vancouver: Talonbooks, 2002).